MILITARY VEHICLES IN PRESERVATION

Royston Morris

First published 2024

Amberley Publishing
The Hill, Stroud,
Gloucestershire, GL5 4EP

www.amberley-books.com

Copyright © Royston Morris, 2024

The right of Royston Morris to be identified as the Author
of this work has been asserted in accordance with the
Copyrights, Designs and Patents Act 1988.

All rights reserved. No part of this book may be reprinted
or reproduced or utilised in any form or by any electronic,
mechanical or other means, now known or hereafter invented,
including photocopying and recording, or in any information
storage or retrieval system, without the permission in writing
from the Publishers.

ISBN: 978 1 3981 1533 0 (print)
ISBN: 978 1 3981 1534 7 (ebook)

British Library Cataloguing in Publication Data.
A catalogue record for this book is available from the British Library.

Typeset in 9pt on 12pt Celeste.
Typesetting by SJmagic DESIGN SERVICES, India.
Printed in the UK.

Contents

	Acknowledgements	4
	Introduction	5
Chapter 1	Tanks	6
Chapter 2	Self Propelled Guns (SPGs)	10
Chapter 3	Large Trucks	13
Chapter 4	Small Trucks	18
Chapter 5	Wheeled Armoured Personnel Carriers (APCs)	22
Chapter 6	Wheeled Armoured Fighting Vehicles (AFVs)	25
Chapter 7	Armoured Cars	28
Chapter 8	Scout Cars and Reconnaissance Vehicles	32
Chapter 9	Tracked Vehicles	36
Chapter 10	Universal Carriers	40
Chapter 11	Half-Track Vehicles	44
Chapter 12	Artillery Tractors and Prime Movers	47
Chapter 13	Jeeps	51
Chapter 14	Light Utility Vehicles	54
Chapter 15	Medium Utility Vehicles	58
Chapter 16	Land Rovers	61
Chapter 17	Specialised Purpose-built Vehicles	65
Chapter 18	Staff and Command Cars	69
Chapter 19	Wreckers and Recovery Vehicles	73
Chapter 20	Emergency Vehicles	76
Chapter 21	Amphibious Vehicles	80
Chapter 22	Off-road 4x4 Vehicles	83
Chapter 23	Motorcycles and Bicycles	86
Chapter 24	The Restoration Process Awaits	90
Chapter 25	Gate Guardians	93

Acknowledgements

The author would like to thank the following organisations for allowing photographs to be taken on their premises and giving their permission for them to be published in this book:

The Cobbaton Combat Collection, Chittlehampton, Umberleigh, Devon, EX37 9RZ
www.cobbatoncombat.co.uk

The Imperial War Museum, Duxford Airfield, Cambridgeshire, CB22 4QR
www.iwm.org.uk

Armourgeddon Military Museum, Southfields Farm, Husbands Bosworth, Leicestershire, LE17 6NW
www.militarymuseum.uk

The Grampian Transport Museum, Montgarrie Road, Alford, Aberdeenshire, AB33 8AE
www.gtm.org.uk

The British Motor Museum, Banbury Road, Gaydon, Warwickshire, CV35 0BJ
www.britishmotormuseum.co.uk

Haynes International Motor Museum, Sparkford, Yeovil, Somerset, BA22 7LH
www.haynesmuseum.org

National Emergency Services Museum, Old Police/Fire Station, West Bar, Sheffield, South Yorkshire, S3 8PT
www.visitnesm.org.uk

Eden Camp Modern History Museum, Malton, North Yorkshire, YO17 6RT
www.edencamp.co.uk

The author would like to thank all those owners and individuals who have preserved and restored vehicles and take them to various shows and rallies to be enjoyed by visitors from all over the world.

Introduction

Before the First World War, armies across the world might only have had a small handful of road vehicles such as a few troop-carrying lorries and the odd tank or two. The hostilities and the build-up to them, however, saw countries expanding their fleets of military vehicles in significant numbers. This increase in vehicles continued throughout the 1920s. With uncertainty regarding whether or not there would be another war during the 1930s, but with all the signs pointing towards this, large-scale orders for more military vehicles were placed with manufacturers to ensure there were enough available.

Following the conclusion of the Second World War in 1945, governments were keen to keep their numbers of military vehicles at a maximum (just to be on the safe side in case there was another war) and this policy lasted into the mid-1950s, when manufacturing of these vehicles slowed dramatically and only a handful were produced. The 1960s saw another change in the role of the military vehicle with conflict in Vietnam and troubles in places such as the Middle East and Northern Ireland. Governments had to have the most sophisticated and up-to-date technological vehicles designed for warfare, so once again manufacturers saw their order books expand to meet the demand. This trend continued upward with more advanced vehicles being produced into the 1970s and carrying on to the present day following military conflicts in the Falkland Islands, Afghanistan and Iraq, to name a few.

In this book I aim to look at the many and diverse types of military vehicles that have retired from service and are now cared for by preservationists and museums alike.

Chapter 1

Tanks

The earliest tanks were built during the First World War in around 1915 and looked nothing like their later successors. Tanks were originally designed as a heavily armoured special weapon to replace armoured cars. They were intended be used for getting across the trenches and going through barbed wire, among other unusual tactical situations.

The earlier tanks were fairly crude and often unreliable, however, during the intervening years between the two world wars, they were developed significantly and eventually became the mainstay vehicles of ground armies. They are classed as armoured combat fighting vehicles and have a turret on the top of the body that rotates 360 degrees and can fire from any position within that circle while the vehicle is in motion.

A First World War Mk IV Male heavy tank replica, built by the 'Lest We Forget Organisation' in 2003. It was seen at the 'Tracks to the Trenches' event held to commemorate the start of the First World War at the Apedale Community Centre, Chesterton, Staffordshire, on 13 September 2014.

A Churchill MkVII Crocodile Flame Thrower infantry tank, built by Vauxhall in 1944. This 44-ton beast has 152-mm-thick body armour, its main weapon is a 75-mm gun and flame thrower and it could tow a 400-gallon flame fuel trailer. It operated with the 79th Armoured Division of the REME. It was seen at the Cobbaton Combat Collection, Devon, on 12 December 2021.

This Centurion A41 Mk13 main battle tank (MBT) served with the Royal Engineers throughout its military career until its retirement in 2008, when it was purchased by the owner of the Armourgeddon Military Museum at Husbands Bosworth, Leicestershire. It arrived there in 2009 and is seen on 23 June 2012. Weighing in at 51 tons, with armour ranging from 51-mm to 152-mm thick, its main armament came in the form of a 105-mm L7 rifled gun (the inside of the barrel is spiralled), backed up with a .30-calibre Browning machine gun.

The AMX-13 light tank, built by Atelier de construction de Roanne, weighed 14.5 tons and had armour between 10 and 40 mm thick. Its main weapon was a 105-mm canon with secondary weapons comprising a single 7.62-mm machine gun and two smoke-grenade dischargers. Used by the French Army, this example from the 1960s was located at the now-closed Fort Paull Museum near Hull, where it was photographed on 17 October 2017. Its turret was replaced with a mild sheet-steel fabrication for an episode of the television programme *Dempsey and Makepeace*.

Sherman tanks came in a number of different varieties. This example, seen at the Capel Military Show in Surrey on 2 July 2022, is a 1945-built M4A2E8 variant which served with the 66th Armoured Regiment in the Omaha Beach D-Day landings. It came fitted with a 76-mm gun and a .50- and .30-calibre machine gun, weighing 33 tons with 50.8-mm armour.

Leyland produced the 'Centaur MkIV' cruiser tank. It was armed with a 95-mm six-pounder howitzer gun. This example, built in 1944, was seen at Cobbaton on 12 December 2021, carrying Royal Marine Armoured Support Group markings. It weighs 28 tons with 64-mm armour on the hull and 76.7 mm on the turret front. In 1945, it was converted to a bulldozer tank by the Army and sold for scrap in 1957; it was acquired in 1978.

This Russian T34/85 medium tank was built by the Kharkov Locomotive Factory in 1944. It weighs 29 tons with body armour ranging from between 15- and 60-mm thick in various places. The main weapon comes in the form of a 76-mm F-34 tank gun, which is backed up by two secondary weapons, 7.62-mm machine guns. It was imported from the Czech Army Reserves to Cobbaton in 1989, where it was seen on display on 12 December 2021.

An M3A5 Grant Command medium tank, built at Baldwin Locomotive Works in 1942. It weighs 27 tons with 51-mm armour at the hull front, turret front, sides and rear with 38-mm thick armour at the hull sides and rear. It had a Browning machine gun as a secondary weapon. This example was used by General Montgomery during both battles at El Alamein in 1942. In order to make room for extra communications equipment its 37-mm gun main weapon was replaced by a wooden dummy. It was seen at the Imperial War Museum, Duxford, on 23 September 2023.

Chapter 2

Self Propelled Guns (SPGs)

Most Self Propelled Guns look almost identical to tanks, but that is really where the similarity ends. The vast majority of these vehicles were built on a tank chassis with various body styles available. They were not as heavily armoured as tanks and as a result not as likely to survive in direct-fire combat. The other main difference is that the gun cannot be fired while the vehicle is in motion and has to be stationary with the gun barrel in a locked-on position.

Built in 1943 at the Montreal Locomotive Works, Canada, this twenty-five-pounder Sexton MkII Self Propelled Gun (SPG) was built onto a Ram tank chassis. It never saw active service in the Second World War, and in 1946 was sent on contract to the Portuguese Army. When fully loaded with 112 shells and a tank of fuel, it weighed around 25 tons. It remained in Portugal until the mid-1980s when it was purchased by a private collector. It was seen at Cobbaton on 12 December 2021.

This Sd kfz 173 Jagdpanther Anti-Tank SPG was built by MIAG in 1944. It weighs 50 tons and has body armour ranging from 40 mm at the rear to 80 mm at the front. It is armed with an 88-mm gun and has a 7.92-mm machine gun as a secondary weapon. The 'waffle' effect seen on the vehicle is Zimmerit, a paste applied all-over to prevent magnetic mines from sticking to it. It is pictured at the Imperial War Museum, Duxford, on 23 September 2023.

The FV433 Field Artillery SPG was built by Vickers Ltd and designed as a replacement for the Sexton. Used by both the British and Indian armies since the 1960s, this vehicle comprises a 105-mm field gun and has 10-mm and 12-mm armour throughout. These vehicles were named 'Abbot', this particular example having been built in 1963 and on display at the Gloucestershire Steam Extravaganza, Kemble Airfield, on 7 August 2010.

This 1943 SPG was built by the Nuffield Organisation. The Morris C9B is based on a Morris C8 Artillery Tractor chassis. This was lengthened and re-enforced with heavy steel plate, four jacks lock the platform carrying the 40-mm Bofors Anti-Tank gun into place and it has armoured shields fitted to the sides and rear. It served with the Devonshire Regiment, which was part of the 15th Indian Division in Burma, and in 1995 came to Cobbaton, where it is seen on 12 December 2021.

A 1959-built Czechoslovakian State Arsenal M53/59 Praga Anti-Aircraft SPG. It weighs 10 tons, has an aluminium body armour of 10 mm and its main weapon is a 30-mm twin autocannon. It was used by the Serbian forces in Bosnia and then abandoned. In 1999, British troops found it hidden in a barn and it was bought back to the UK. It was on display at the Imperial War Museum, Duxford, on 23 September 2023.

In 1992, Vickers designed the Howitzer Artillery System SPG, known as the AS-90, These entered service in 1993. It can fire 155-mm high explosive between 15.5 and 19 miles, depending on whether the long or short barrel is employed. The body armour is all welded steel with a maximum 17-mm thickness. In January 2023, British Prime Minister Rishi Sunak announced that along with fourteen Challenger II tanks, thirty AS-90s and 45,000 rounds of ammunition would be sent to Ukraine to assist in the ongoing war with Russia. This example of an AS-90 was on display at the Capel Show on 2 July 2022.

CHAPTER 3

Large Trucks

General Service Vehicles (GSVs)

These vehicles were the most numerous of the trucks that were used by armies across the globe. They were used to transport the vast majority of goods from food supplies to equipment, as required by troops in the field.

Introduced into service with the US Army Ordnance Corps in 1941 was the GMC CCKW 2.5-ton 6x6 Off Road Truck. These came in two types, with a variety of body styles. This 1942 hard-cab version was the short wheelbase (145 in) Model 352 fitted with a winch. Here it is taking part in the Maiden Newton 'At War Weekend' event, which was held in the Dorset town over 23 and 24 June 2012. Note the GMC lettering at the top of the front grille.

13

Pictured on 2 July 2022 at Capel is this (164 in) long wheelbase Model 353 canvas cab with no winch version GSV. There were two cab types available, the hard and canvas. Note the lack of GMC lettering on the front grille and the cut-down doors on the cab, which is prevalent on all canvas-cab-type vehicles.

The truck pictured here at the Overlord Show in Denmead, Hampshire, on 5 June 2022 is often seen on Britain's roads. It is a Rheinmetall MAN MV HX60 18.330 4x4 General Service Winch Truck, which was built in 2013 and retired from the British Army in 2019. It was purchased at a military vehicles auction by the present owner in 2021.

Ammunition Carriers

These vehicles by the very nature of the loads they carry are purpose-built, with a composite wood (mainly ash) and steel body with either a 5- or 7-ton underfloor well, which could be lowered to ease loading and unloading, along with shell carriers attached to rails in the floor of the bed of the lorry to make it even safer.

This Matador was an ammunition and troop carrier, built in 1940 by the Associated Equipment Company. It was released from service with the British Army in 1967 and was then used as a timber crane until the early 1990s. When photographed on 5 September 2010 it was taking part in the GDSF.

Troop Carriers

These were usually General Service Vehicles (GSVs) with certain modifications added to carry army personnel.

The Bedford RL saw steady use with the British Army. This 1964 modified example has had a hole drilled into the cab roof and a ring attached, which was used to mount a machine gun or searchlight, making it an armoured vehicle. It was seen on 18 June 2010 at the Lister Tyndale Steam Rally held at Berkeley Castle, Gloucestershire. Note the mount is in the unusual position of being in the centre of the cab.

Radio Trucks and Mobile Workshops

These vehicles have a normal truck chassis with specially built bodies. They acted as communications centres or mobile workshops, the latter mainly repairing guns and other small pieces of equipment used by soldiers while on the move.

Introduced to the British Army in 1970, the Bedford MK 4x4 proved very successful. This 1983 example is a 4-ton Radio Truck and was seen after ending its Army service days in the late 1990s on display at the Great Dorset Steam Fair (GDSF), Tarrant Hinton, on 5 September 2010.

The REO Motor Company represented the US Army in significant numbers. This 1958 M35 'Whistler' 2.5-ton 6x6 Cargo, with a repair shop body, was pictured at the Abbey Hill Steam Rally, Yeovil, on 30 April 2022. The term 'Whistler' comes from the unusual noise that emits from the axles when the vehicle is in motion at certain speeds.

NAAFI Staff Canteens

These vehicles were used to supply army personnel with the food and drink needed for them to fulfil their duties to the best of their ability.

This long-wheelbase GMC CCKW353 6x6 NAAFI Vehicle was built in 1943 and used by the US Army Ordnance Corps until the mid-1960s. It was then sold for scrap but rescued from a scrapyard and restored. It was seen on 2 July 2022 at the Capel Show, where it was masquerading as an American Red Cross Clubmobile.

Chapter 4

Small Trucks

These numerous small or lightweight trucks were used by various armies and military organisations throughout the world when a large truck was deemed unnecessary for the task in hand.

In 1969, Steyr-Daimler-Puch developed the 'Pinzgauer' range of small high-mobility all-terrain trucks for use primarily by the Austrian Army. These trucks came in two styles. This example, pictured at the Capel Show on 2 July 2022, is a 710 model 4x4 GSV, which was built in 1974 and formerly saw use with the Swiss Army.

This 1974-built 'Pinzgauer' is the larger 6x6 712M model GSV FFR (Fitted For Radio) and was seen on display at the Overlord Show on 5 June 2022. The name 'Pinzgauer' comes from a domestic breed of cattle hailing from the Salzburg region of Austria, where they are at home in any terrain ranging from fields to mountains.

In 1980, the second-generation 'Pinzgauer' was introduced. The main differences were the restyled front end and change to the designations. This 716 version (formerly 710) was on display at the Capel Show on 2 July 2022.

The multi-purpose 'Unimog' truck was manufactured by Daimler-Benz from 1951 onwards and sold under the Mercedes-Benz brand. The early versions, such as this 1951 example, were given the designation 2010 and sold under the Unimog brand. They didn't bear the traditional Mercedes-Benz three-pointed star badge (which appeared from 1953 onwards). This vehicle was seen at Abbey Hill on 30 April 2022 and had been in service with the Swiss Army until its retirement.

This 1978 Unimog is a Type 421 Cabrio and was used by the Royal Marines for over twenty-five years before being pensioned off and purchased by a private buyer. It was soaking up the hot spring sunshine at the Basingstoke Festival of Transport on 8 May 2022.

Developed in 1978 and given the designation of RB44 (originally RB510), these multi-purpose 4x4 light utility trucks were built by Reynolds Boughton specifically for use by the British Army. They entered service in 1989 and following an announcement in 2010 that they were being decommissioned, several enthusiasts purchased one and these can be seen at military shows. This 1993 example with a mobile-workshop body attached is seen at Capel on 2 July 2022.

The Volkswagen Transporter was first introduced in 1949 and is now in its seventh generation. The (Type 2) T3 was introduced in 1979 and is one of the more popular generations. This 1988 version served with the German Feldjager (Military Police) until it was decommissioned and offered for sale in 2009. It was seen at Capel on 2 July 2022.

The Dodge WC series of military trucks were light and medium utility vehicles given the nickname 'Beeps'. The series came in many different designations, ranging from WC1 to WC64. This 1941 example, a WC12 ½-ton pick-up truck, was on display at Capel on 2 July 2022.

Chapter 5

Wheeled Armoured Personnel Carriers (APCs)

These comprise a broad type and range of armoured vehicles that were designed to transport personnel and equipment to combat zones. They have less armament than similar looking vehicles and were not designed to provide direct-fire support in battle. They were also designed as self-defence vehicles and not specifically built to fight on their own.

A FV603 Saracen Armoured Personnel Carrier, built by Alvis, seen on 5 June 2022 at Denmead. Its main weapon was a Browning M1919 .30-calibre machine gun. This 1954 example was decommissioned in the early 2000s following service with the British Army in the Iraq War and was purchased by the owner at an auction.

A Humber FV1611 (designated Pig) lightly armoured truck, which was used primarily by the Royal Ulster Constabulary (RUC) from 1958 until early 1970s. It was a prominent feature on the streets of Northern Ireland during this period. This retired 1955 Mk2 version was on display at the South West England Festival of Transport, held in Yeovil, on 11 August 1991.

The Ford M20 was a conversion of the M8 Greyhound Armoured Car. The conversion involved removing the turret and replacing it with a low armoured top open superstructure, but retaining the M6 37-mm gun as its main weapon and fitted with a machine-gun ring for a .50-calibre M2 heavy machine gun. This 1943 built version was on display at the Capel Show on 2 July 2022.

Designed in 2006 as an infantry mobility vehicle for the US Army, this is an International MXT-MV truck, built by Navistar Defense. In 2009 the British government ordered 262, which were modified to comply with MOD requirements for Tactical Support Vehicles, known as Husky TSVs. This vehicle on display at the IWM, Duxford, on 23 September 2023, shows how well protected the main vehicle was following an attack from an IED (Improvised Explosive Device).

A 1953 example of an ex-RUC Humber Pig looking resplendent in the sunshine at Capel on 2 July 2022. It is in non-authentic but strikingly colourful livery.

Built in 1956 by Alvis as a Mk2 FV603 Saracen, this vehicle entered service in British Malaya during the Malayan Emergency. It remained there until 1961, when it was returned to England, heavily modified and changed to a Mk5, for use in Northern Ireland. It stayed there until retired from service in 1986, when it was donated to the Tank Museum, Bovington. It is seen here in its role as a gate guardian to the museum's archive and library collections centre on 21 January 2022.

Chapter 6

Wheeled Armoured Fighting Vehicles (AFVs)

The main type of vehicle that comes under this heading is the Infantry Fighting Vehicle (IFV), an armoured fighting vehicle used to carry infantry troops into battle and also to provide direct-fire support. They were designed to be more mobile than tanks and are equipped with a rapid-firing autocannon or a large conventional gun and can also have side ports for infantry soldiers to fire their small arms out at the enemy.

A FV601 Saladin Armoured Fighting Vehicle, designed by Crossley Motors of Manchester in 1954. They were subsequentially built by Alvis. On display at the Overlord Show on 5 June 2022 was this 1958 model fitted with a 76-mm gun and a Browning machine gun. It has 32-mm-thick body armour and weighs a hefty 11.5 tons. It served with the Royal Signals Regiment.

The history of this Saladin is unknown at the time of writing, only that it was built in 1959. It was formerly used as a gate guardian of the 15th/19th Hussars at their Fenham Barracks HQ in Newcastle upon Tyne before moving to the North East Land, Sea & Air Museum (NELSAM) in Sunderland, where it was pictured on 19 October 2014.

The Marmon-Herrington Armoured Car was produced in South Africa. Weighing 6.4 tons, its main armament was a QF two-pounder gun with a Browning machine gun as secondary armament and it had a 20-mm-thick armoured body shell. These vehicles mainly saw use in North Africa, with a vast number taking part in the seven-month Siege of Tobruk in 1941. This example was built in the early part of 1941 and was sent straight to the above-mentioned action with the 1st King's Dragoon Guards. It was pictured at Fort Paull Museum on 17 October 2017.

A 1943-built M8 Greyhound, pictured at Armourgeddon, Leicestershire, on 23 June 2012. It was retired from service with the US 83rd Infantry Division. It has 25-mm-thick armour and a M6 37-mm gun as its main armament. Note the rearmost flat tyre.

Weighing in at 7.5 tons is the Daimler Armoured Car, introduced into the British Army during the early 1940s. It was produced in tandem with the Dingo Reconnaissance Scout Car. The main armament of these vehicles was a QF two-pounder gun, along with a 7.92-mm machine gun and body armour up to 16-mm thick in places. This 1942 MkI version was photogaphed on 12 December 2021 at Cobbaton. This vehicle was used in the 1977 film *A Bridge Too Far*.

The AML H-90 is a lightweight armoured fighting vehicle – weighing 5.5 tons, built by Panhard. Its major feature was the DEFA D921 90-mm low-pressure rifled canon and had 60-mm mortar to back it up. It was also armed with a 7.62-mm coaxial machine gun. Pictured at Fort Paull on 17 October 2017, this 1962 version was used by the Argentine Army. It was abandoned and subsequentially captured during the Falklands War in 1982 and then bought to the UK for display at the museum.

Chapter 7

Armoured Cars

This chapter covers those vehicles which, although armoured, were specifically designed to perform subordinate battlefield tasks such as internal security, armed escort and passive observation, to name a few.

In September 1914, all available Rolls-Royce Silver Ghost chassis were requisitioned by the government to form the basis for a new type of Armoured Car. It would have 12-mm-thick body armour, a fully rotating turret and a .303 machine gun; when completed the cars would weigh 4.7 tons. It was known as '1920 and 1924 Pattern' for those modified in the corresponding years. This '1924 Pattern' was on display at the Apedale 'Tracks to the Trenches' event on 13 September 2014.

Timoney Technology Ltd built this 4x4 MkIV Armoured Car for the Irish Army Cavalry Corps in 1979. It weighs just under 10 tons and has 12.7-mm body armour and 9.5-mm-thick underside armour. It was equipped with 2 x 7.62-mm machine guns. Following its retirement from service, it was donated to the Howth Castle National Transport Museum of Ireland, where it was pictured outside on 3 April 2015.

One of the interesting vehicles seen at Capel on 2 July 2022 was this 2020-built replica of a 1918 German Freikorps Revolution Armoured Car. It was built during the Covid-19 pandemic in the UK by a retired engineer, seventy-six-year-old John Atkinson. It is a mild-steel vehicle built onto the chassis of an old Morris car saloon and has an authentic-looking autocannon and a genuine 7.62-mm machine gun.

The BTR-60 is an eight-wheeled Armoured Car made by GAZ in Russia. This example, seen at the IWM, Duxford, on 23 September 2023, dates from 1960. It weighs 11.5 tons and has welded steel body armour between 6 and 9 mm thick on the hull and between 7 and 10 mm on the turret. Its main weapon comes in the form of a 14.5-mm heavy machine gun backed up with a 7.62 coaxial machine gun. It was used by the Warsaw Pact in the Cold War during the 1960s and 1970s.

In 1940, Lord Beaverbrook instructed Standard Motor Company to build a series of Armoured Cars to be used by the Army. The result was the Beaverette, which had 11-mm armour backed by 3-inch-thick oak planks. They were open at the top and rear and had a .303 machine gun. Later versions, like this MkIII, had a turret fixed to the top. The hull was redesigned giving the wings and front end a flat, sloping look. This 1941 example, built onto the chassis of a Standard Flying Fourteen saloon, served with the Irish Army and upon its retirement from service ended up at Cobbaton. It was pictured there on 12 December 2021.

The Staghound Armoured Car was built by the Chevrolet Motor Company during the Second World War. This 1944 T117E1 version, seen at the IWM, Duxford, on 23 September 2023, served with the Free Belgian Forces from new, and was acquired by the museum in 2002. It was armed with a 37-mm gun, a .30-calibre Browning machine gun and a 2-inch smoke mortar in the turret, with the hull containing a second .30-calibre Browning. These 14-ton vehicles had body armour that was an astounding 44 mm thick.

The Shorland Armoured Car was built by Short Brothers in Northern Ireland onto the 109-inch Land Rover Series IIA chassis especially for the RUC and the Ulster Defence Regiment. The armour was designed to withstand 7.62-mm bullets and the main armament was a 51-mm general purpose machine gun. This 1967-built prototype was seen at the Collections Centre at the British Motor Museum, Gaydon, on 10 June 2023.

Chapter 8

Scout Cars and Reconnaissance Vehicles

These vehicles were usually lightly armoured and only carried small or low-calibre weapons, which would have been used in close infantry combat only if necessary.

The White Motor Company built the M3A1 Scout Car. These vehicles weighed 6.25 tons and had 13-mm-thick body armour. They were equipped with a .50-calibre Browning machine gun with 2 x .30-calibre Browning machine guns as secondary weapons. This example was built in 1941 and was used by the British Army until the early 1980s. It was pictured at Cobbaton on 12 December 2021. The round cylinder shape on the front is an 'unditching roller', which enabled it to cross a 1.5-foot-wide trench with no difficulties.

The Sd.Kfz.222 is a German Light Armoured Reconnaissance vehicle built by Auto Union. These 4-ton vehicles had 14.5-mm-thick armour and were armed with a 20-mm Autocanon with a secondary recoil-operated air-cooled general purpose machine gun. This 1942-built example saw action in the North African Campaign and was captured during the hostilities. It was brought to the UK by a private collector. It was pictured at the Buckinghamshire Railway Centre during a 'Military Vehicles at War' event on 15 August 2010.

The 3-ton Daimler 'Dingo' Armoured Reconnaissance vehicle was the forerunner to the 'Ferret'. These vehicles had 30-mm-thick armour at the front and 12-mm-thick armour on the sides. They were armed with .303 Bren light machine gun or a .55 anti-tank rifle. This 1942-built Mk2 example, seen at Cobbaton on 12 December 2021, saw service with the 6th New Zealand Infantry Brigade and took part in the battles at El Alamein.

Daimler built the 'Ferret' Armoured Scout Car mainly for use with the British Army, and it remained in service until 1992. It weighed 3.7 tons and had body armour ranging from 6 to 16 mm thick. This 1953 example is a FV701 Mk2 variant and would have been fitted with 7.62-mm general purpose machine gun and a .30 Browning machine gun. Seen at Capel on 2 July 2022, bearing the colours of the 10th Royal Hussars, with whom it served in 1956/57, this vehicle served with a total of eight different regiments over thirty-eight years.

A 'Ferret' pictured at the Tank Museum, Bovington, on 21 January 2022. One of only fifty 'Ferrets' built to the Mk5 variant and designated FV712, it was built in 1967 and is equipped with a 'Swingfire' turret, which is fitted with Anti-Tank Guided Missiles that could engage and destroy an enemy tank at a range of 4,000 metres.

The Fox FV721 4x4 Combat Reconnaissance Vehicle was built by Royal Ordnance in Leeds for use with the British, Malawi and Nigerian armies. It comprised an all-welded aluminium armoured body, 33.5-mm thick at the front, 41-mm thick on the sides and 38-mm thick at the rear, and weighed 6.75 tons. Its main armament was a 30-mm Rarden Canon, with a co-axial 7.62-mm machine gun as its secondary weapon. This 1977 example was spotted on a low loader parked at the roadside in the Herefordshire village of Stretton Sugwas on 16 September 2007.

Built by GKN Sankey Ltd in 1983, this Saxon AT-105 Internal Security Patrol Reconnaissance Vehicle was used by the British Army. It featured heavily in the Afghanistan conflict between 2007 and 2010 before coming to the IWM at Duxford, seen there on 23 September 2023. It is built onto a revised Bedford 'M' 4x4 chassis and weighs 11.7 tons. Its hull is welded steel with a 'V'-shaped under-chassis plate to deflect mine detonations. It is armed with a 7.62-mm machine gun and is protected further by a steel cage surrounding it.

Chapter 9

Tracked Vehicles

All fully tracked vehicles, be they tanks, SPGs, AFVs or the vehicles covered in this chapter, require specialised training for personnel to be permitted to drive them because of the complex steering mechanisms incorporated within them.

Not all military vehicles with tracks are used in battle. Some were built as gun, equipment or personnel carriers and some were used in Arctic locations or desert terrain.

The FV432 'Trojan' is an Armoured Personnel Carrier built by GKN Sankey Ltd. Weighing 15 tons, with 12.7-mm-thick body armour, the main weapon was a 7.62-mm general purpose machine gun, with secondary armament in the form of smoke dischargers. This example was built in 1961 and served with UN peacekeeping forces in the Middle East. It was seen at Cobbaton on 12 December 2021.

The Allis-Chalmers M4 High Speed Artillery Tractor was in service with the US Army from 1943 until 1960. These vehicles were used to tow a 90-mm anti-aircraft gun or a 155-mm gun or 8-inch howitzer. They came equipped with a M2 Browning machine gun and weighed 14.2 tons. Upon withdrawal from service in the early 1960s, this 1944 version was used as a rock-drill carrier for a road construction company in British Columbia. It was on display at Capel on 2 July 2022.

The Schutzenpanzer Lang HS.30 was a West German infantry fighting vehicle, designed and built by Hispano-Suiza. An initial contract was placed for 10,680 vehicles, but following a series of technical issues, the order was cut back to 2,176, which cost the West German government 40 million DM in compensation. The vehicle came in eight different variants. This 1960-built example, seen at Fort Paull on 17 October 2017, is a SPz Typ 52-3 self-propelled mortar version with a 120-mm Brandt mortar fitted.

Combat Vehicle Reconnaissance (Tracked) (CVR(T))

These vehicles were designed specifically by Alvis and were available in half a dozen or so different variants. They were generally small, highly mobile, air-transportable armoured vehicles.

The 'Sabre' (designated FV101A) was a hybrid vehicle of the CVR(T) vehicle series combining the hull of a FV101 'Scorpion' and the turret of a Fox reconnaissance vehicle. The main armament was a 30-mm cannon along with a 7.62-mm chain gun. This 1971 version was taking part in the Overlord Show on 5 June 2022.

The FV102 'Striker' in the CVR(T) series entered service in 1976 with the Royal Artillery of the British Army of the Rhine, before being transferred to the Royal Armoured Corps. This variant was armed with the 'Swingfire' wire-guided anti-tank missile system, which could hit a target effectively and accurately in a range of between 150 and 4,000 metres. Its secondary armament was a 7.62-mm general purpose machine gun. This 1973-built example was seen at Denmead on 5 June 2022.

Armoured Vehicle Royal Engineers (AVRE)

These vehicles formed a group of armoured engineering vehicles that were used to protect engineers during front-line battlefield operations.

By protecting the engineers, the vehicles became mobile platforms for a variety of engineering purposes, ranging from mounting large-calibre weapons for demolition to carrying engineering stores and mine-clearing explosives, and also for deploying roadways and modified bridges.

The FV434 is the Armoured Repair Vehicle variant of the FV432 APC, introduced into service during the early 1960s. It was primarily used for changing the Chieftan tank power packs in the field by the Royal Electrical & Mechanical Engineers (REME), who still use these vehicles today. This 1969 GKN Sankey Ltd-built version was purchased from the MOD in 1992. It had formerly been used during exercises at BATUS (British Army Training Unit Suffield) in Canada and was pictured at Cobbaton on 12 December 2021.

All-Terrain Vehicles

These vehicles work best equally travelling on snow, sand, mud and swamp terrains.

Devoloped in the 1950s by a Canadian inventor, the Flectrac Nodwell FN22L tracked ATV troop carrier was designed and used by the Canadian Army. They were lightweight weighing 2.8 tons, and they came in two track sizes of 28 and 36 inches. This undated example was seen at Fort Paull on 17 October 2017.

Aktiv Maskin Ostersund Ltd specialised in making the 'Snow-Trac', a vehicle about the size of a modern compact car. It came in two different versions, with either a two- or seven-person cab and a hard or canvas top. The version seen here is a ST4B, built in 1974 and sold to the Irish Army, with which it served until retirement in the mid-1990s. Purchased by the Irish Turf Board for use on their peat bogs, it proved to be not up to the job. It is pictured out of use in a compound at the Boora Works in County Offaly on 16 May 2023, awaiting a decision on its future.

Chapter 10

Universal Carriers

These light armoured tracked vehicles were widely used by British Commonwealth forces during the Second World War, usually for transporting personnel and equipment (mostly support weapons) or even as mobile machine-gun platforms.

Engineers Captain Vivian Loyd and 6th Baronet Sir John Carden formed the Carden-Loyd Tractor Ltd company in the 1920s. This was purchased by Vickers-Armstrong Ltd following the death of Sir John in 1935. Loyd left and formed the Vivian Loyd Company when his relationship with Vickers deteriorated. Loyd's speciality was Universal Carriers. This 1942 MkI version is known as a Towing Tractor [TT]. It has 7-mm body armour and weighs 4.5 tons. Seen at the Overlord Show on 5 June 2022, it would have been used to tow a six-pounder 57-mm anti-tank gun.

Eden Camp Modern History Museum in Malton, North Yorkshire, is a fascinating place to visit, as I did on 28 August 2021. Among the military vehicles on display was this 1943 American-built T-16 Mk1.

A MkII, built by the Vivian Loyd Company in 1943. It is pictured in the Land Warfare Hall at the IWM, Duxford, on 23 September 2023.

This 1944-built Vickers-Armstrong Weapons Carrier served with the Royal Artillery Anti-Tank Regiment during the Second World War. Its post-war history is not known. It was seen on 4 August 2012 at the Claude Jesset Great Bush Railway, Hadlow Down, during a 'Military Vehicles Day'.

The most numerous variant carrier produced was the TT version, and this 1942 example saw service with the 7th Armoured Division ('Desert Rats') during the North African Campaign. It was built by the Wolseley Motor Company, based on a Loyd design. It was seen on display at Cobbaton on 12 December 2021.

This Bren Gun Carrier No. 2 MkII was built in March 1944 by Ford Canada. It saw military service with the 43rd Infantry Division. Pictured on 5 June 2022 at Denmead, it weighs 3.5 tons and has 10-mm body armour and a Bren light machine gun as its main weapon.

The Ford Canada 'Windsor' TT versions were the late comers of the Second World War. These 5-ton vehicles had body armour of 9.5 mm on the front and 6.3 mm on the sides. However, the floor was the weak point as it only had a 3.1-mm steel plate, which left the vehicle susceptible to mine damage. This 1944 example has a new canvas top fitted with rolled up sections that can be used as viewing/driving window spaces. It was pictured at Cobbaton on 12 December 2021.

Chapter 11

Half-Track Vehicles

These vehicles have regular wheels at the front and continuous tracks at the rear end of the vehicle. The idea behind this design was to provide a vehicle with the cross-country capabilities of a tank and the handling of a wheeled vehicle.

The main advantage these vehicles have over wheeled vehicles is that the tracks reduce the pressure on a given area of the ground by spreading the vehicle's weight over a larger area. They also do not require the complex steering mechanisms found on tracked vehicles and anybody with a driving licence can drive one without the need for speciality training.

International Harvester began producing these M5 Half-Track vehicles in 1941. They have 15.8-mm armour and a 13-mm machine gun backed up with 2 x 7.62-mm machine guns. This 1942 M5 was just one of many vehicles that took part in the 'At War Weekend' in Maiden Newton on 23 and 24 June 2012.

This 1943-built White M3 T19 Half-Track is pictured taking part in the Overlord Show on 5 June 2022. This version is a gun carriage with a 105-mm howitzer. Its military history is not known to the author.

PSD, a Czechoslovakian engineering company, built Tatra OT-810 Half-Track APCs for the Czechoslovakian Army. These vehicles weighed 9 tons, were armed with a 7.62-mm machine gun, had 15-mm frontal armour and 8 mm on the sides, and were renowned for being a copy of the German Sd.Kfz 251. This 1960-built example in the guise of a Sd.Kfz 251/7.1 Pionierpanzerwagen, engineer's assault vehicle fitted with fixtures to carry assault bridge ramps, was taking part in the Denmead Show on 5 June 2022.

Another Tatra OT-810 that took part in the Denmead Show on 5 June 2022 was this 1961-built example, which was in the guise of a German Wurfrahmen 40 multiple rocket launcher. Both of the Tatras on display during the show looked very authentic and could easily be mistaken for the German vehicles that they represented.

Built in 1962, this Tatra OT-810 was used by the Czechoslovakian Army from new until it was retired from service in 1984; its subsequent history is not known by the author. It was pictured at the Eden Camp Museum, Malton, on 28 August 2021.

A 1961-built Tatra OT-810 in the Land Warfare Hall, IWM, Duxford, on 23 September 2023.

The Black Cruise expedition was organised by Citroën and took place between 1924 and 1925, covering 12,427 miles from Algeria to Madagascar. Eight Citroën Autochenille Half-Track vehicles took part. This 1922 example was one of the eight and is pictured at the Victoria & Albert Museum, London, as part of the 'Cars – Accelerating the Modern World' exhibition on 22 February 2020. It was acquired by the US Army in 1927 as part of their evaluation of half-track vehicles. After finalising these evaluations, it was donated to the National Car & Tourism Museum in Compiègne, France.

Chapter 12

Artillery Tractors and Prime Movers

Artillery Tractors

Artillery tractors (sometimes referred to as gun tractors) are specially built tractor units designed to tow artillery pieces of various weights and calibres. There are two main types: 'wheeled' and 'tracked'. Wheeled vehicles are usually purpose-built lorries modified for military use and tracked vehicles are often built on modified tank chassis with the superstructure replaced with a compartment to hold the gun crew or ammunition.

The Morris Commercial C8 FAT (Field Artillery Tractor) is instantly recognisable by its 'pug-nosed' front end. It weighed 3.3 tons, its sloped sides indicating that it was sufficiently armoured, although, surprisingly, it had no body armour whatsoever. This 1939-built MkII version served with 10 Corps in the battles at El Alamein. It is pictured at Cobbaton on 12 December 2021, with its twenty-five-pounder howitzer gun just visible to the right at the rear.

The Four Wheel Drive Company (FWD), formed in 1909, produced this 6-ton 1942 truck. A COE (Cab Over Engine) variant, it was initially used by the British Army as a gun tractor and retired from service in 1946. It was acquired from MOD sales in 1947 and then spent fifty years as a mobile crane for a timber haulage firm. It was seen on 5 September 2010 at the GDSF.

This 1950 Scammell Explorer 6x6 was the eighth vehicle off the production line and was new to the REME 29th Southern Command Workshop. It saw service there until the 1980s, when it was bought by a private collector. It is pictured taking part in the GDSF on 3 September 2010.

Canadian Military Pattern (CMP) trucks were a range of 3-ton 4x4 trucks built by General Motors, Ford and Chrysler of Canada. Over half a million examples and versions were built to British Army specifications. Built in 1944 by Ford Canada, this F60S LAAT (Light Anti-Aircraft Tractor) served with the 1st Canadian Army until 1946, when it was used by the National Bus Company as a recovery truck until the 1960s. It is seen here at Cobbaton on 12 December 2021.

Prime Movers

These purpose-built, heavy duty towing tractor units were used to haul heavy and awkward loads such as tanks, aircraft and other often oversized objects.

This 1940 Mack NM 6-ton 6x6 Prime Mover retired from the US Army as a tank recovery tractor in 1949 and was then purchased by Hibble & Mellors Funfair of Nottingham to haul rides. In 1962, Aberdeen County Council acquired it and converted it into a snowplough. In 1977, ownership passed to Grampian Regional Council, which still used it as a snowplough until 1985, when it was donated to the Grampian Transport Museum in Alford. It is seen on display there on 31 July 2022.

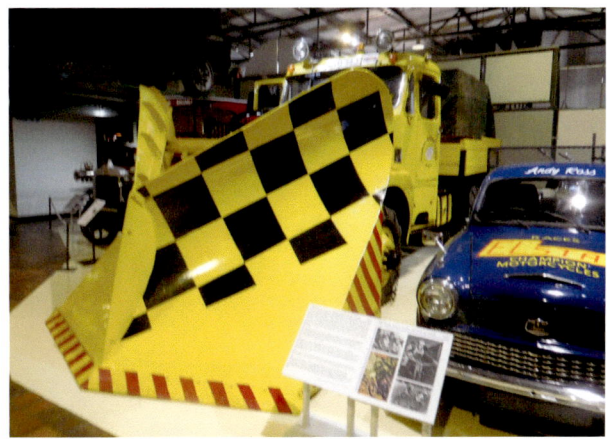

This 1967 Kaiser-Jeep-built M52 5-ton 6x6 Prime Mover has the capacity to tow a 16.7-ton load, but on improved roads this is increased to 24.6 tons. August Bank Holiday weekend 2012 played host to the Honiton Hill Rally, Devon, where this former US Marine Corps vehicle was pictured.

The Scammell Commander is a heavy equipment transporter that entered service in 1984. It weighs 19.9 tons and can carry loads up to 65 tons. This 1983-built example served with the Royal Corps of Transport in Belgium until the early 2000s. It was seen taking part in the Basingstoke Festival of Transport, ironically carrying the vehicle it replaced – a 1959 Thornycroft Antar tank transporter – on 8 May 2022.

Chapter 13

Jeeps

Out of the hundreds of different light utility vehicles used by armies around the world over the years, the ¼-ton 4x4 Command Reconnaissance Light Utility vehicle, more commonly referred to as the jeep, has been the most numerous by far. Over 600,000 of these vehicles have been built especially for military use.

There were two main producers of jeeps, namely Willys (MB) and Ford (GPW). They weighed just a little over a ton and could travel for 300 miles on a full tank of fuel. Built in 1943 by Ford, this vehicle was used by the Special Air Services (SAS) in North Africa during the Second World War. The differences between this and a normal jeep include the removal of the windscreen to avoid glare, removal of the bars on the grille and a water condenser unit fitted, which captures the steam from the radiator and turns it back to water for reuse. The numerous jerry cans are full of water and fuel, which could give an extended range of 1,250 miles. This one was taking part in the Capel Show on 2 July 2022.

Built in 1944, this Willys is one of the few examples that have had body armour and a 7.62-mm machine gun fitted. It saw service with the RUC on the streets of Northern Ireland in the 1970s. It is seen taking part in the Capel Show on 2 July 2022.

Although Ford and Willys built the majority of jeeps, French car manufacturers Hotchkiss produced the M201 version for use by the French Army. They didn't begin making these until 1955 and they were used until 2000. They were known by the French military as LATV (Light All-Terrain Vehicle). This 1957-built version was taking part in the Capel Show on 2 July 2022.

Early-build Willys M38s had narrow metal strips in their slatted grilles, as seen on this 1942 example at Denmead on 5 June 2022.

In late 1953, Willys was the subject of a takeover by Kaiser-Jeep, who manufactured the M38A1 until 1971. Following a complete overhaul, this 1961-built example was being offered for sale for £8,500 at Capel on 2 July 2022.

In 1961, Ford was awarded the contract to replace the aging Willys MD and M38 models. They came up with the Ford Mutt M151, which to all intents and purposes was virtually identical to the Second World War models, the main difference being the front grille. Willys having the copyright to use the vertical slotted version, Ford used a horizontal slotted one. This had five slots, as can be seen on this 1963 example on display at Overlord on 5 June 2022.

The contract to build the M151 was also later awarded to Kaiser-Jeep and AM General. Although their versions were based on the Ford design, there were minor differences. The AM General version, known as the M15A2, had a front grille with five horizontal slots in line with Ford, except that the top two slots were shorter than the remaining three. This 1971 AM General version was seen at Capel on 2 July 2022.

Chapter 14

Light Utility Vehicles

Apart from the previously mentioned jeeps, the military have used several different types of light vehicles even though the majority of them are used to perform the same job.

The 4x2 Light Utility Cars, or 'Tillys' as they were known, were British Army utility vehicles based on mid-sized saloon cars. The transformation included putting a pick-up body with a canvas top on the car chassis. This 1944 Morris 10 version was seen at the GDSF on 5 September 2010. The present owner has had it in their possession since 1977 and formerly it served with the 43rd Infantry Division.

Another 1944-built 'Tilly' is this Standard Flying 12 hp. It weighs 1.3 tons and bears its original WD number. It served with the South West Southern Command Bomb Disposal Squad. Following the war, it was used by an Exeter builder until 1963 before being purchased by the Cobbaton Collection, where it was seen on 12 December 2021.

In the late 1940s, a specification for a British Army light field truck was issued to replace the American jeeps. The contract was awarded to the Morris Car Company, which came up with the 'Nuffield Gutty', but sadly it was not the success that had been hoped for. This 1947-built ¼-ton prototype was seen on display at the British Motor Museum, Gaydon, on 10 June 2023.

Originally known as 'Truck ¼ton CT 4x4 Cargo FFW Austin Mk1', the 'Austin Champ' was built by the Austin Motor Company for the British Army. The military designation was FV1801A and these 12-feet-long vehicles had a top speed of 64 mph. This 1953-built example was taking part in the Denmead Show on 5 June 2022.

55

The 'Mini Moke' was an airportable light utility vehicle introduced in the British Army in the 1960s. Built by various manufacturers, it weighed just under ½ ton. This 1964 BMC version was pictured hanging from the ceiling at the Haynes International Motor Museum, Sparkford, on 10 August 2013. It is believed to be one of the only surviving BMC military versions in the UK.

Between 1946 and 1951 Škoda Auto AS built the 1101A light military vehicle for use by the Czechoslovakian Army. It was 13 feet 2 inches long and weighed just over 1 ton. This 1947 example is a radio car. It was found abandoned in Holland as part of a large collection of Škoda vehicles in 2014 and was seen on display at Capel on 2 July 2022.

Russian manufacturer GAZ produced the GAZ 67 for the Russian military. The vehicle was marketed as 'the Soviet equivalent of the Jeep'. In 1944, it was upgraded to the GAZ 67-B, which weighed 1.2 tons. This 1945-built 67-B was found in a barn in Latvia in 2011 and brought to the UK by a collector. Following its two-year restoration, it was used in George Clooney's 2014 film *The Monuments Men*. It was seen at Capel on 2 July 2022.

In 1953, GAZ was asked to replace the 67-B and came up with a four-wheel-drive light off-road vehicle, the GAZ 69. Production ended in 1956, when the contract was cancelled and transferred to the UAZ Automobile Plant. As UAZ specialised in the production of four-wheel-drive vehicles, it was felt that they had better facilities to do the job. The UAZ vehicles were known as the 69M. They were 12 feet 6 inches long and weighed 1.6 tons. This 1966-built 69M was on display at Capel Show on 2 July 2022.

Chapter 15

Medium Utility Vehicles

These vehicles come in various forms ranging from water bowsers to weapons carriers.

The Dodge WC series of military vehicles comprised various models, the most numerous of these being the WC51 and WC52 Weapons carriers. Both types of these vehicles were designated by the US Army as ¾-ton 4x4 Truck (G-502). This 1943 WC51 former 12th Cavalry Division example weighs 2.6 tons and was on show during a public open day held at Joe Nemeth Engineering Ltd at Easter Compton on 2 October 2010.

The main and most noticeable difference between the WC51 and the WC52 is that the WC52s have a winch attached to the front of the radiator just beneath the grille. This 1944-built WC52 weighs 2.8 tons and was taking part in the Annual New Year's Day Road Run, held by the Somerset Traction Engine Club in Taunton, Somerset, on 1 January 1999.

The Dodge T-215 ½-ton 4x4 truck, known as the WC22, was the forerunner to the previously mentioned vehicles. It weighed 2.4 tons. This 1941-built example was taking part in the Capel Show on 2 July 2022.

The Bedford MW 4x2 15 cwt light truck, weighing in at 2.4 tons, was another mainstay of the British Army. This 1942-built example is an MWC and fitted with a 200-gallon water bowser. It served under the Second Tactical Air Force, 83 Squadron, and featured in the Battle of Arnhem. It was pictured on 23 September 2023 at the IWM, Duxford.

In the 1950s, the Dutch Army replaced their vehicles. In December 1951, DAF was given 175 million guilders by the Dutch government to supply a series of military trucks. One of these was the YA126 light utility truck. It weighed 3.3 tons and had a 310-mile range. This 1956 example of a weapons carrier was decommissioned in 1996 and was seen at Abbey Hill on 30 April 2022.

The AM General-built High Mobility Multipurpose Wheeled Vehicle (HMMWV), 'The Humvee', is a light military truck. This 1995 example is an M1114 Expanded Capacity Vehicle (ECV) Armament Carrier. It provided protection for the occupants from 7.62-mm armour-piercing projectiles, 155-mm artillery air bursts and 12-lb anti-tank mine blasts and weighed 6 tons. In December 2014, the American Department of Defense began auctioning off Humvees that were regarded as surplus. This vehicle, seen at Capel on 2 July 2022, was one of those.

The 4x4 M715 light vehicle was designed in 1965 by Kaiser-Jeep to replace the Dodge M37. It weighed 3.7 tons and offered a 225-mile range. They were primarily used in the Vietnam War, although the US Army considered them underpowered and fragile compared with the vehicles they replaced. This example, seen at Capel on 2 July 2022, was built in 1967 and operated with the 7th Cavalry Headquarters Division.

Chapter 16

Land Rovers

The Land Rover is one of the most reliable and sturdy off-road vehicles that has ever been produced. It is, therefore, only natural that the military would use it and commission many different variants to their own designs to maximise the key features.

Pictured on 5 June 2022 at Denmead, this 1951 Series I 80 Inch Land Rover was used as an Aircraft Crash Rescue Vehicle with the F11 Fighter Command at RAF Tangmere, near Chichester. It remained in service there until August 1958, when it was purchased at an auction in Lincolnshire.

In the 1960s, the Royal Marines and the British Army were looking to replace the Austin Champ and they approached Land Rover to manufacture that vehicle. The smallest available at that time was the Series IIA 88 Inch, which proved to be too heavy, so Land Rover modified this by applying parts that could be easily removed or by not fitting certain parts at all and the result was the ½-ton Lightweight Airportable. This 1971-built version was used as a radio communications unit and was seen at the Overlord Show on 5 June 2022.

The Land Rover 101 Forward Control (FC) was a light utility vehicle that was produced solely for use with the British Army. Designed to be airportable and for use as gun tractors, they were 14 feet 2 inches long, weighed 2 tons and had a 311-mile range. They were not available to the general public off the production line but were as military surplus, and there are several now in private ownership. This 1975-built gun tractor and ammunitions carrier was seen at the Somerset Steam Spectacular Show at Low Ham on 17 July 2010.

In the early 1990s, Land Rover converted thirty-one surplus 101FC vehicles for use in the 1995 futuristic film *Judge Dredd*, starring Sylvester Stallone. This example was converted in 1994 (build year not known) and was seen on display at the British Motor Museum, Gaydon, on 10 June 2023.

In 1968, the MOD bought seventy-two Series IIA Land Rover 109-inch vehicles, and Marshalls of Cambridge converted them for specific use by the SAS. This included the fitting of four fuel tanks, giving the vehicles a range of 1,000 miles, making the chassis and suspension heavy duty, removing the doors, windscreen and roof, and fitting two 7.62-mm machine guns. The SAS painted them pink and named them 'Pink Panthers', the reason for this being it camouflaged the vehicles in the desert. They were used by the SAS until 1984 and then retired. Of the seventy-two, about twenty are known to have survived. This 1969-built example, seen at Denmead on 5 June 2022, is one of those.

In March 1983, JRA (Jaguar-Rover-Australia) Limited was given a contract to supply the Australian Army with a new light military utility vehicle. The 'Perenti' was based on the Land Rover 110. This 1991 4x4 version weighs 2.5 tons, has twin fuel tanks and room to carry fourteen fuel/water carriers. In 2013, these vehicles began to be sold off. This one is a Long Range Surveillance Vehicle, which was used for five- to six-month long-range trips to watch coastal areas for smugglers and illegal immigrants landing. It is one of two in the UK and was imported in 2016. It is seen here taking part in the Capel Show on 2 July 2022.

After the Second World War, the Belgian firm Société Anonyme Minerva Motors was awarded a contract to build their own version of the Land Rover 80 Inch under licence for use by the Belgian Army. This was known as the Minerva TT (Tout Terrain) 4x4, weighed 1.2 tons and was fitted with a 7.62-mm machine gun. This 1952-built version, seen on display at Capel on 2 July 2022, served until 1990 and was then purchased by a private buyer. Note the sloping front wings, which clearly define it as being a Minerva vehicle.

Chapter 17

Specialised Purpose-built Vehicles

Almost any army in the world has vehicles within their numbers that have been built for at least one specific purpose and as such are used solely for the job they were built to do. This often means that the vehicle is uneconomical to keep as it is it parked up more than it is utilised. Often due to the nature of the job they were designed to do, these vehicles can be difficult to modify for other uses, although, if they could be, the cost of such modifications is usually far too expensive to contemplate carrying out.

This 1944-built GMC CCKW352 former US Ordnance Corps truck was built as a Pontoon Bridge Carrier. It was used to transport temporary bridges to locations where they would be built and used for the purpose they were needed and then removed afterwards and conveyed to another location. It was seen at the GDSF on 3 September 2010.

The Sd.kfz 302 Goliath Light Charge Carrier comprised a series of unmanned German vehicles that were loaded with just under 1 ton of explosives. Built by Borgward, they weighed just under ½ ton, had 5 mm of body armour and were remote-control operated. This meant they could be guided up to an Allied tank and then exploded, destroying both tank and vehicle. This 1943 example was seen at the IWM, Duxford, on 23 September 2023.

This 1962-built Citroën 2CV was converted by the French Army for use in the Algerian War. It is an airportable mobile Gun Platform and equipped with a 20-mm Mauser MG 151 gun. It is seen here at the Castle Combe Steam & Vintage Rally on 19 May 2013.

The Royal Signals and Radar Establishment was a scientific research establishment within the MoD base in Malvern. Technologies such as radar, satellite communications and defence communications were just a few areas that were researched and developed at the facility. This 1980-built Bedford M1120 was specially constructed and fitted out according to the correct specifications. It was seen at Kemble Airfield, Gloucestershire, on 7 August 2010.

The Artillery Saturation Rocket System (ASTROS) is a self-propelled multiple rocket launcher made by the Avibras Company in Brazil. This 10-ton beast can carry thirty-two 13-feet 9-inch-long 127 mm to 450 mm missiles with a firing range of 18.5 miles. It also had a 12.7-mm Browning machine gun as back-up weapon. This 1985 version was seen at the IWM, Duxford, on 23 September 2023.

The Russian 9K33 Osa SA-8 Gecko 6x6 amphibious low-altitude short-range surface-to-air missile system, built by Znamya Truda, was introduced in 1971. It weighed 17.5 tons, had a ground clearance of a mere 400 mm and a firing range of 7.5 miles. This 1980 version, seen at the IWM, Duxford, on 23 September 2023, is an upgraded 9K33M3, which had a firing range of 9 miles.

In use with the US Army, the M752 self-propelled rocket launcher transported the MGM-52 'Lance' surface-to-surface tactical ballistic missile system. The system was developed at the height of the Cold War and produced by LTV in Dallas, Texas. It weighed between 1.4 and 1.6 tons (depending on the warhead carried). These systems would carry a thermonuclear warhead that had a blast yield of 1,000 tons of TNT, which travelled at a greater speed than Mach 3 (2,301 mph), with a range of 81 miles. They were in service for twenty years between 1972 and 1992. This example was seen at Capel on 2 July 2022.

67

00 ET 20 is a 1970-built Foden 6x4 26-ton Armoured Riot Control Water Cannon that was used by the RUC in Northern Ireland in the 1970s. It was fitted with a Pyrene Water Cannon and is believed to be the only one in existence. It was pictured at Fort Paull on 17 October 2017.

Photographed at the Tank Museum on 21 January 2022, this Royal Ordnance FV4201 Chieftain MBT is fitted with a mine plough. This device was intended to clear a path through a minefield, buried mines being ploughed up and either pushed outside of the tank's path or turned upside down, which meant that when they were detonated the blast was a downward one and therefore would lessen the damage to the vehicle and its crew.

Chapter 18

Staff and Command Cars

Staff Cars

The function of these vehicles is self-explanatory. They were used by senior ranking officers to travel around visiting the troops at garrisons and regimental quarters, as well as by the enemy making trips to POW camps.

This 1938-built Hillman Minx Saloon was one of 500 that were purchased by the Ministry of Supplies in 1938/39 for use as ministry staff cars. This example carries the markings of the 3rd Infantry Division, with whom it served. It was pictured on 5 September 2010 at the GDSF.

This 1960-built Standard Ensign Estate staff car of the RAF Fighter Command No. 12 Group was seen at the GDSF on 5 September 2010. The vehicle operated at RAF Watnall, Nottinghamshire.

This 1942-built Studebaker President Saloon staff car was pictured at Denmead on 5 June 2022. Its military history is unknown to the author.

Built between 1940 and 1945, a total of 50,500 Volkswagen Type 82 Kubelwagens (bucket-seat car) were used by the German Army mainly as a light liaison vehicle, although it was also used as a staff car. These 4x2 no-frills, easily and cheaply built vehicles weighed 0.7 tons. This 1944 version staff car was one of two taking part at the Capel Show on 2 July 2022.

Command Cars

These vehicles are more often light utility vehicles, which were kitted out with the appropriate equipment required. They were used by senior officers as a mobile command post to keep an eye on how things were going in a battle, but not put those men in a position of any danger.

Another vehicle in the Dodge WC series was the WC56 Command Car. This 1942-built example was seen at Capel on 2 July 2022 and was used by the US 8th Army Air Force. Following the end of the Second World War, it was sold at a disposal auction in Dorchester in 1945 and used for the next three years. Then it stood unused in a barn until 1967, when it was purchased and restored.

This 1941-built Humber 4x4 was originally made as a pick-up truck. It was converted to an officer's truck by the REME in 1945 to enable it to be used by the 3rd Infantry Division. It is seen here taking part in the GDSF on 5 September 2010.

The G-Class series of 4x4 off-road vehicles was introduced by Mercedes-Benz in 1979. This 1980 example of the military version, namely the 230-G Jeep, was captured by the REME during the Falkands War with Argentina. It was the command vehicle that was used by Brigadier General Menendez. After being shipped to the UK in 1985, it now resides at Cobbaton and was seen there on 12 December 2021.

Between 1953 and 1958, Volvo built the TP21 four-wheel-drive off-road vehicles for use by the Swedish and Belgian armies. These 2.5-ton vehicles had 3.6-litre engines. They have been given the nickname 'The Suggan' by enthusiasts. This 1957 example was on display at the Capel Show on 2 July 2022.

Chapter 19

Wreckers and Recovery Vehicles

Just as in civilian life, the military relies on vehicles that can be called upon to rescue and recover stricken vehicles when they are in trouble. It was not practical to use conventional civilian recovery vehicles, so the military created their own fleet of usually purpose-built vehicles.

This 10-ton 6x6 M1 Heavy Wrecking Truck was built by Ward LaFrance in 1943, especially for the US Army. Weighing 13.6 tons and fitted with a 9-ton winch on the front, it could tow vehicles up to 29.5 tons. It was on display at Joe Nemeth Engineering on 2 October 2010.

Built by Leyland in 1958, this 10-ton 6x6 Martian Heavy Wrecker saw service with the REME in the Cyrenaica District of Eastern Libya until the 1970s before being repatriated to the UK. It then saw further use at the Defence School of Electronic & Electrical Engineering at MoD Lyneham in Wiltshire until the early 2000s, when it was deemed surplus to requirements. It was pictured on display at the GDSF on 5 September 2010.

This AEC Militant MkIII heavy duty recovery vehicle was built for the British Army in April 1970. It didn't see active service until October 1971, when it was sent to Northern Ireland to serve with the REME Light Aid Detachment, which was attached to the 16th/5th Queen's Royal Lancers. It was deemed surplus in 1987 and sold at an Army auction and its history over the next twenty years is sparse. The current owner purchased it from a plant dealer in Scunthorpe in 2007 and spent two years restoring it. It was seen at Capel on 2 July 2022.

Built by Austin in 1941, this K6 6x4 3-ton recovery vehicle served with the REME attached to the 3rd Armoured Division during the Second World War. Following the end of its military service, it was preserved by the REME 325 Advance Workshops Long Range Recovery Group and was pictured taking part in the South West England Festival of Transport at Yeovil on 11 August 1991.

A 1942-built Scammell 30-ton TRMU Tank Recovery Tractor, seen at Cobbaton on 12 December 2021. This vehicle served with the XXX Corps in the Western Desert and was used to remove stricken tanks during the Second Battle of El Alamein in late 1942.

A 1952 Austin K4 recovery truck, photographed at the Claude Jessett Trust's Great Bush Railway, Hadlow Down, during a military vehicles day on 4 August 2012. This vehicle formerly served with the King's Own Scottish Borderers at their base in Berwick-upon-Tweed.

This 1941 recovery truck was built by the Diamond 'T' Company, who between 1905 and 1967 manufactured trucks and automobiles, prior to merging with REO trucks in 1968. It was pictured on 24 June 2012 at the Maiden Newton 'At War Weekend'.

Chapter 20

Emergency Vehicles

Ambulances

As in the previous chapter, armies don't rely on civilian emergency vehicles but have their own fleets of these vehicles. The most popular vehicles used as ambulances included those built by Austin, Dodge and Land Rover.

A 1943-built Bedford OYD Mobile X-Ray truck, seen at Kemble Airfield on 7 August 2010. It was used by the British Army until the early 1970s.

Built in 1941, this Dodge WC54 ¾-ton 4x4 Ambulance was taking part in the Maiden Newton 'At War Weekend' event on 24 June 2012. It served with the French Liberation Army during the final years of the Second World War. Following the end of hostilities, it was acquired by a private collector and bought to England, where it frequently attends various military shows.

The Canadian Army used this 1944-built Chevrolet C15TA Armoured Ambulance from new until the mid-1970s. It was then used by a mountain-rescue team as a field ambulance until the early 1990s. It was then purchased and taken to Cobbaton, where it was pictured on 12 December 2021. Note the snow chains on the tyres.

Auxilliary Fire Service (AFS)

The AFS was formed in 1938 and comprised unpaid part-time volunteers, whose role as part of the Civil Defence Service was to supplement the work of fire brigades at local level. The organisation was replaced in 1941 by the National Fire Service (NFS), which initially used the vehicles that were in service with the AFS but were rebranded with the new organisation's logo.

Auxiliary Fire Service (AFS) vehicles were painted in a light-grey all-over livery. The main vehicles used were Austin K2 trucks in the form of auxiliary towing vehicles. This 1941-built unbranded example was pictured at the National Emergency Services Museum, Sheffield, on 30 August 2021. Note the blackout headlight covers.

In August 1941, the AFS became the National Fire Service (NFS). Looking resplendent in its NFS red livery under an overcast sky, this 1943 Austin K2 Fire Engine was taking part in the Somerset Steam Spectacular Show on 19 July 2009.

Airfield Fire Vehicles

These vehicles were used at RAF bases throughout the country.

Built in 1964 by Thornycroft, this Nubian Major 6x6 Mk9 500-gallon Foam Tender was seen in Market Harborough, Leicestershire, on 23 June 2012. Prior to it being purchased for preservation, this vehicle spent the majority of its service at RAF Coningsby, Lincolnshire, and is now under the care of the Museum of RAF Firefighting.

Civil Defence Corps

In 1949, the NFS was incorporated into a new civilian volunteer organisation to form the Civil Defence Corps. Its purpose was to take local control of an affected area in the aftermath of a major national emergency. The organisation was disbanded in 1968, however, in 1956, there were still 330,000 personnel involved in it.

The RLHZ Self Propelled Pump, 'Green Goddess', built by Bedford, was used by the Civil Defence Corps until 1967, when it was mothballed but retained by the government in case of a nuclear emergency, the threat from the Soviet Union still prominent at that time. They were used by the Army most notably during the firefighters strike in the late 1970s. Deemed as surplus in 2004, numerous examples were sold off at auctions. This 1955 example was used by the 825 Naval Air Squadron at Yeovilton Airfield, Somerset, and was pictured at the West Somerset Railway Steam Fayre on 1 August 2009.

Chapter 21

Amphibious Vehicles

The military has certain vehicles that are not only adept at travelling on land but are equally as capable of doing the job travelling in water.

Vickers-Armstrong built this 1942 Valentine DD Duplex MkIX amphibious tank. These tanks worked by erecting a canvas 'flotation screen', which would enable them to float in water and was lowered upon reaching land. This example served with the 9th Armoured Brigade during the Second World War. After the war ended, it was purchased by a farmer, who used it as a bulldozer before selling it to the current owner. It was seen taking part in the Overlord Show on 5 June 2022. Note the propeller between the tracks, which was used to move the tank forward while in the water.

In 1942, General Motors introduced the DUKW, which was a six-wheel-drive amphibious vehicle based on the modified chassis of their CCKW trucks and referred to as 'Duck'. It was used by the US military during the Second World War and the Korean War. This vehicle, like the DD tank, had a single propeller at the rear. This example is seen on display at the Capel Show on 2 July 2022.

The FV622 was classified by the British Army as a HMLC (High Mobility Load Carrier) 5-ton 6x6 Amphibious Alvis Stalwart. It weighed 9.5 tons and had a 400-mile range. It was used by the British Army until 1993 and then classed as surplus, this 1964 example being one of the many that were bought at auction after this. It was pictured at Capel on 2 July 2022.

The Polish People's Republic built the OT-64 8x8 Amphibious Personnel Carrier. Weighing 14.5 tons with 123-mm-thick body armour, it had a top speed of 58 mph (but only 5.5 mph in water). This version, built in 1968, is an ex-Polish Army SKOT R-3M unarmed combat engineers signals and mobile command headquarters. It was seen at the West Somerset Railway Steam Fayre on 1 August 2015.

In 1960, the FMC Corporation developed the 12-ton M113 amphibious Armoured Personnel Carrier for the US Army. Originally designed as a lightly armoured airportable battlefield taxi, and used in the Vietnam War, this vehicle had an aluminium-alloy hull which was between 28 and 44 mm thick. It was armed with a M2 Browning machine gun with an automatic grenade launcher as back-up. This 1961 example was seen at the Capel Show on 2 July 2022.

Looking at this 19-ton vehicle you would not think it was amphibious. Built by the Royal Ordnance Factory in Leeds for use by Royal Engineers, the FV180 Combat Engineers Tractor was used by the British, Indian and Singapore armies; it was replaced from 2013 onwards. Its body armour was a honeycombed twin-skin aluminium alloy and its top land speed was 35 mph (9.2 mph in water). This 1978 version was taking a rest between demonstrations at the GDSF on 5 September 2010.

The amphibious version of the Ford GPW Jeep was the Ford GPA Seep (Sea Jeep). These vehicles did not perform as well as the DUKW, mainly due to the production vehicle being ½ ton heavier than specified. Its volume had not been increased accordingly, which led to it getting stuck in shallow waters and making it too heavy on land. This 1943-built example was seen at Capel on 2 July 2022.

Chapter 22

Off-road 4x4 Vehicles

The vehicles covered in this chapter are all off-road 4x4 types. These vehicles were usually deployed in desert areas and in rugged mountainous terrain and were commonly purpose-built.

A 1963-built Fardier FL500, a ½-ton 4x4 light general-purpose vehicle. Developed by Lohr Industrie for French airborne troops, it had a top speed of 22 mph. It is pictured at the Lister Tyndale Steam Rally held at Berkeley Castle, Gloucestershire, on 18 June 2010.

This Fardier FL501, built in 1979 and seen at Capel on 2 July 2022, is different to the example in the previous picture (although it looks the same). The difference is within the engine because it has the capacity to tow a load up to 0.8 tons, which usually comes in the form of a 120-mm mortar gun.

A Rolba Goblin all-terrain 4x4, built in 1984 by the Swedish company Rolba Svenska AB for use by the Swedish Army. This vehicle is a very rare find because only twelve of them were ever built. This example was seen on 5 September 2010 at the GDSF.

The ALSV (Advanced Light Strike Vehicle) was developed by Chenowth Racing in 1996 as an all-terrain 4x4 light vehicle for the American military. This 1.7-ton vehicle was armed with a 12.7-mm machine gun and a 40-mm grenade launcher. It was used extensively in Afghanistan and during the Iraq War and is still in use today. This example was part of a display at the Overlord Show on 5 June 2022.

Tomcar was founded in Israel in 1991 as a manufacturer of off-road 4x4 utility vehicles. The Springer 4x4 is one of the vehicles they designed specifically for use by military organisations. This 2009-built example was formerly in service with the US 16th Air Assault Brigade and is seen here taking part in the Capel Show on 2 July 2022.

The EPS Springer is the British version built by Enhanced Protection Systems and has notable differences in comparison with the Israeli vehicle. These include solid three-quarter-height doors and a more ventilated cab roof. This 2006 version was used by the British Army in Afghanistan and Iraq and was seen taking part in the Capel Show on 2 July 2022.

Chapter 23

Motorcycles and Bicycles

Motorcycles

Most of the modes of transport featured in this chapter were primarily used by the military for sending dispatches and instructions. However, some motorcycle and sidecar combinations were fitted with a machine gun in the sidecar for close fighting or had a stretcher fitted to take the wounded off the battlefield.

The Excelsior Welbike folding motorcycle was the smallest motorcycle used by the British Army. It was designed to fit into a CLE canister, which was 51 inches long, 15 inches high and 12 inches wide. The Welbike were used extensively by the 1st and 6th Airborne Divisions. This 1942-built Series 2 MkII was seen suspended on display in the Airborne Assault Museum of the Parachute Regiment and Airborne Forces, which can be found inside of the AirSpace hangar of the IWM, Duxford, on 23 September 2023.

The Army Medical Service Corps (AMSC), a non-combat speciality branch of the German Army, provided medical cover for the Army during armed conflicts. This 1943 BMW R71 and Sidecar Combination was used by them during the Second World War and was captured by Allied forces on the outskirts of Berlin at the tail end of the campaign. It was at Kemble Airfield on 7 August 2010. Note the stretcher mounted onto the sidecar. The seating area beneath would have contained bandages and medications, etc.

Royal Enfield motorcycles were widely used by the British Armed Forces during the Second World War. This 1943 WDCO version was one of 2,826 that were built as part of RAF Contract C/14219 and whose initial destination was the Sheffield War Office. In 1945, after the war, it was shipped to Central/Eastern Europe to assist in the UNRRA (United Nations Relief & Rehabilitation Administration) aid missions. It was repatriated back to the UK in 2021, renovated and photographed at Capel on 2 July 2022.

This 1942-built KS750 with a BW38 sidecar was produced by the German motorcycle manufacturer Zündapp exclusively for the unified armed forces of Nazi Germany, the Wehrmacht. It saw action in North Africa among other places during the Second World War. It was seen at Capel Show on 2 July 2022.

Bicycles

Bicycles were primarily used by members of the Local Defence Volunteers (LDV, later renamed the Home Guard) and by Air Raid Precaution (ARP) wardens to carry out their duties reasonably quickly and without the need for petrol, which was, of course, highly rationed.

The MO-05 was used by the Swiss Bicycle Infantry (SBI). The most noticeable elements are the large document and map carry-case fitted to the frame, the bag fitted onto the handlebars to hold the rider's battle-helmet and other small items, and, between the frame and the rear wheel, the small-arms gun-carrying case, and a canvas bag fitted on one side of the rear porter rack, which would have held the rider's food rations. This example, seen at Denmead on 5 June 2022, carries a plate featuring the Swiss flag and the date 2001, the year that the SBI was phased out.

This 1940s Raleigh Gents Bicycle was used by an Air Raid Warden (ARP), who was based in Post No. 7 in Wimborne, Dorset, during the Second World War. The large box on the porter rack at the rear would have held his helmet and gas mask, while the bag on the crossbar would have contained important documents, his torch and food rations. It was seen on display at the Overlord Show on 5 June 2022.

Columbia Manufacturing Incorporated in Westfield, Massachusetts, was one of the biggest bicycle manufacturers in America and supplied the US Army with thousands of vehicles. This 1940 example was one such vehicle, a Columbia Westfield 92L folding bicycle. It was seen attached to the front of a Willys MB Jeep at the Overlord Show on 5 June 2022.

Chapter 24

The Restoration Process Awaits

Military museums and private collectors of military vehicles will at certain times have a vehicle or two (or more) that is awaiting restoration. These will often include rusting hulks of metal waiting their turn to be transformed into pristine working or static exhibits on display for everybody to enjoy for generations to come.

A 1945-built former 11th Armoured Division Leyland A34 Comet Cruiser Tank. Pictured at the rear of the Cobbaton Collection on 12 December 2021, it is patiently awaiting its turn to be restored so that it can hopefully one day join the others on display inside the collection.

This unknown build-year Morris C8 Artillery Gun Tractor, pictured during a farm dispersal sale at Elmstone Hardwicke, Gloucestershire, on 24 August 2012, is definitely a project in need of lots of restoration work to bring it back to its former glory.

The Tank Museum at Bovington has a host of vehicles awaiting restoration and the following series of pictures, taken on 21 January 2022, shows some of them in various conditions. In spring 1943, a total of fifty-two Sherman M4A2 tanks were converted into BARV (Beach Armoured Recovery Vehicles) in preparation for the D-Day landings. These vehicles could operate at a depth of 9 feet and could pull loads weighing 17.5 tons on wet sand.

An A22D Infantry Gun Carrier, mounted with an A4 anti-aircraft gun and based on the hull of a Churchill tank. The prototype, fitted with a 3-inch AA gun built by Vauxhall in 1941 and tested in 1942 on Salisbury Plain, was considered satisfactory. The contract was awarded to Beyer, Peacock & Co. in Manchester to build forty-nine of them. By the end of 1942 they were declared obsolete following the introduction of the A30 Challenger tank. This example was used on the Lydd Gun Ranges in Kent (the bullet and shrapnel holes bear testament to that) following the war and was rescued and bought to Bovington in the 1990s.

Built by Nuffield Limited, the A24 Cavalier cruiser tank replaced the A15 Crusader tank. This 1943 example is one of only two survivors. It was used by the 9th Armoured Division as a training tank during the war before becoming a hard target at Larkhill on Salisbury Plain, from where it was saved in 1985. The other survivor is behind this one in the picture.

Awaiting restoration beside the Land Warfare Hall at the IWM, Duxford, is this 1980 UMZ-built Russian ZSU-23/4 'Shilka' Lightly Armoured Radar Guided Anti Aircraft SPG. This 19-ton vehicle was equipped with 4 x 23-mm autocannons as its main armament, with body armour ranging from 9.2 mm to 15 mm thick. This example was captured by the Allied forces during Operation Desert Storm in 1991 and was pictured on 23 September 2023.

Another Russian vehicle awaiting restoration at the IWM, Duxford, is this 1988 GAZ-built BRDM-2 Armoured Amphibious Reconnaissance Vehicle. It weighs 7.7 tons, has body armour ranging from 2 mm to 14 mm thick in various places, was armed with a 14.5-mm heavy machine gun and backed up by a 7.62-mm coaxial general purpose machine gun. This example, pictured on 23 September 2023, was abandoned after the Lebanese Civil War in 1990 and then brought to the UK.

Chapter 25

Gate Guardians

At the gateways and entrances to numerous military establishments up and down the country vehicles (usually tanks and occasionally planes) that have been plinthed can be seen. They are often referred to as 'gate guardians'. Although not strictly regarded as preserved, these vehicles have usually been decommissioned from service and given the role of gate guardian.

This GKN Sankey Ltd FV510 Warrior Light Tank is located at the entrance to the Royal Armoured Corps Gunnery School in Lulworth, Dorset, and was pictured on 14 June 2012.

There are several 'gate guardians' dotted around the Tank Museum at Bovington Camp and the surrounding area. Many of the following series of photographs of these were taken on 21 January 2022. Located outside the main entrance to the museum itself is this 1979-built Royal Ordnance Challenger I Prototype tank.

Located on King George V Road opposite the museum are two gate guardians, the most prominent of which is this 1984-built Royal Ordnance FV4030/4 Challenger 1 MBT. It is located outside the Royal Ordnance Corps (RAC) Stanley Barracks of the Junior Leaders Regiment.

This 1999 BAE Systems-built Challenger II was lifted into place at the entrance to the National Army Museum, Lambeth Road, Chelsea, on 26 March 2021. This vehicle, used by the Royal Armoured Corps in the Kosovo, Bosnia and Iraq conflicts, was pictured on 3 June 2022.

Built in 1944, this A34 Comet Cruiser tank is positioned at the entrance to the IWM, Duxford. This vehicle took part in the Berlin Victory Parade of July 1945. After the war it was at the Army School of Artillery in Larkhill, Wiltshire, until 1970, when it was then donated to the IWM. It is seen here on 23 September 2023.

This Royal Ordnance Factory FV4201 Chieftain tank stands guard outside the MoD Defence Equipment & Support Headquarters on the A46 in Ashchurch, Gloucestershire.

Built in 1943, this Sherman M4A4 Light tank saw action during the D-Day landings with the US 1st Infantry Division on Omaha Beach. It was damaged during that campaign and abandoned as the Allies advanced. After the war it was recovered and repaired by the French Army and used by them for training until the 1980s. Further restoration was then completed in France and in 2016 it was shipped back to the UK to act as a gate guardian for the D-Day Centre Museum in Portland, Dorset, where it was pictured on 2 April 2022.

I end on a poignant note. In April 1944, 30,000 personnel of the US Army and Navy were taking part in 'Exercise Tiger' at Slapton Sands, Devon, a location that bears a resemblance to Utah Beach where they would be landing during D-Day. On 28 April, eight Landing Craft in convoy loaded with troops and equipment were attacked at around 1.30 a.m. by six German E-Boats. Initially thinking it was part of a live-firing exercise, by the time they realised they were under attack the damage had already been done. The outcome was two craft sunk and 749 US personnel dead, many succumbing to hypothermia in the cold water as they awaited rescue. In 1974, local man Ken Small discovered a submerged tank some 60 feet from the shore. After ten years of legal wrangling, it was raised from its watery grave and this 1943-built Sherman M4A1 Duplex Amphibious was that vehicle. It was serving with the US 70th Tank Battalion and now stands in the beach car park as a memorial to those who sadly perished. It was pictured on 18 December 2021.